Matter and Energy Bingo Book

A COMPLETE BINGO GAME IN A BOOK

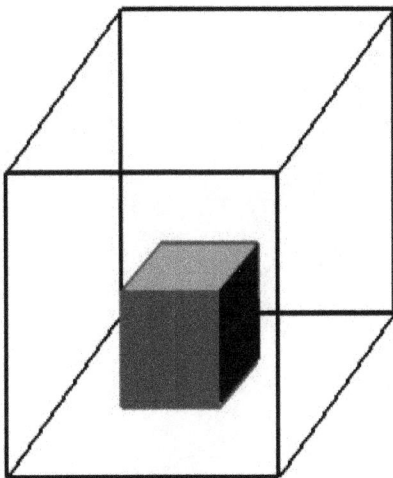

Solid

Holds Shape

Fixed Volume

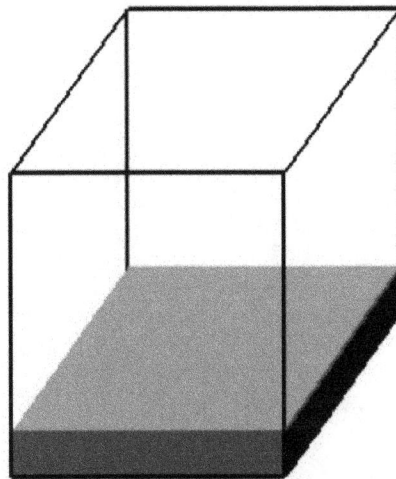

Liquid

Shape of Container

Free Surface

Fixed Volume

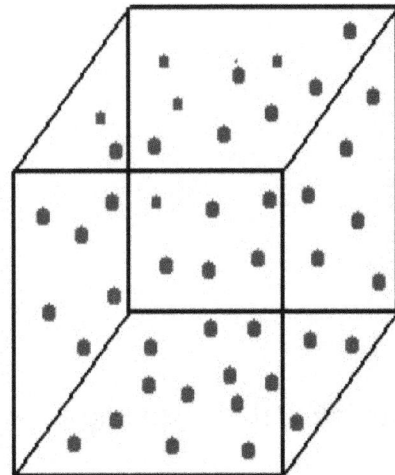

Gas

Shape of Container

Volume of Container

Written By Rebecca Stark

ISBN 978-0-87386-435-0
Educational Books 'n' Bingo

Printed in the U.S.A.

MATTER & ENERGY BINGO DIRECTIONS

INCLUDED:

List of Terms

Templates for Additional Terms and Clues

2 Clues per Term

30 Unique Bingo Cards

Markers

1. **Either cut apart the book or make copies of ALL the sheets. You might want to make an extra copy of the clue sheets to use for introduction and review. Keep the sheets in an envelope for easy reuse.**

2. Cut apart the call cards with terms and clues.

3. Pass out one bingo card per student. There are enough for a class of 30.

4. Pass out markers. You may cut apart the markers included in this book or use any other small items of your choice.

5. Decide whether or not you will require the entire card to be filled. Requiring the entire card to be filled provides a better review. However, if you have a short time to fill, you may prefer to have them do the just the border or some other format. Tell the class before you begin what is required.

6. There are 50 terms. Read the list before you begin. If there are any terms that have not been covered in class, you may want to read to the students the term and clues before you begin.

7. There is a blank space in the middle of each card. You can instruct the students to use it as a free space or you can write in answers to cover terms not included. Of course, in this case you would create your own clues. (Templates provided.)

8. Shuffle the cards and place them in a pile. Two or three clues are provided for each term. If you plan to play the game with the same group more than once, you might want to choose a different clue for each game. If not, you may choose to use more than one clue.

9. Be sure to keep the cards you have used for the present game in a separate pile. When a student calls, "Bingo," he or she will have to verify that the correct answers are on his or her card AND that the markers were placed in response to the proper questions. Pull out the cards that are on the student's card keeping them in the order they were used in the game. Read each clue as it was given and ask the student to identify the correct answer from his or her card.

10. If the student has the correct answers on the card AND has shown that they were marked in response to the *correct questions,* then that student is the winner and the game is over. If the student does not have the correct answers on the card OR he or she marked the answers in response to *the wrong questions,* then the game continues until there is a proper winner.

11. If you want to play again, reshuffle the cards and begin again.

Have fun!

TERMS INCLUDED

atom(s)

attract

Celsius

charge

circuit(s)

compound

condensation

conduction

conductors

current

dissolve(d)

electricity

electromagnet

electrons

element(s)

energy

evaporation

Fahrenheit

frequency

gas

gravity

heat

insulator

kinetic

liquid

magnet(s)

mass

matter

mixture

molecule

neutron(s)

nucleus

opaque

pitch

potential

power

prism

protons

reflection

refraction

repel

solar

solid

solution

sound

temperature

translucent

transparent

wavelength

weight

Additional Terms

Choose as many terms as you would like and write them in the squares.
Repeat each as desired. Cut out the squares and randomly
distribute them to the class.
Instruct the students to place the square on the center space of their card.

Matter and Energy Bingo

Clues for Additional Terms

Write two or three clues for each new term.

_____

1.

2.

3.

_____

1.

2.

3.

_____

1.

2.

3.

_____

1.

2.

3.

_____

1.

2.

3.

_____

1.

2.

3.

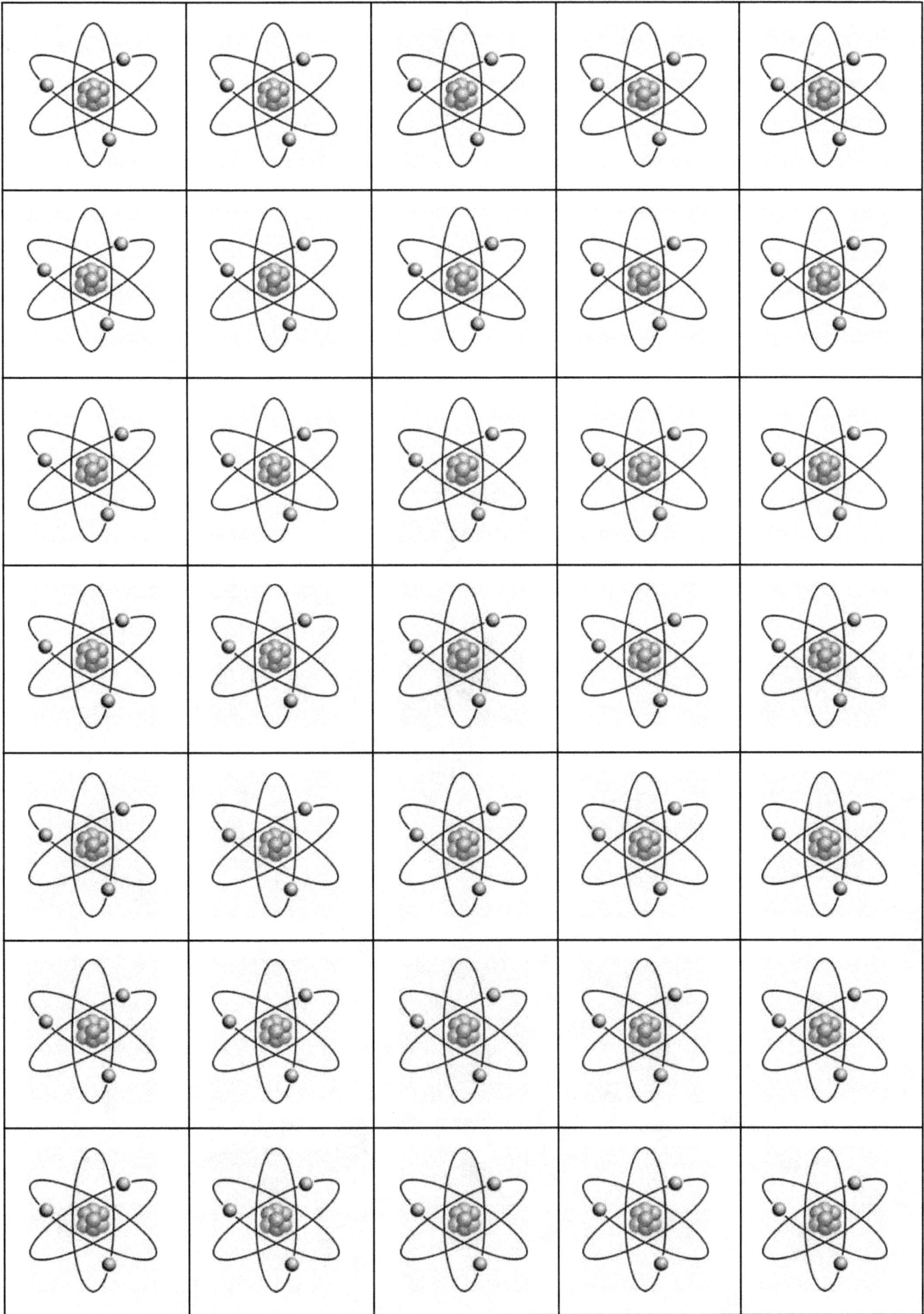

atom(s) 1. Elements are made up of ___. 2. An ___ is composed of protons, neutrons, and electrons. It cannot be broken down by chemical means.	**attract** 1. Unlike poles of a magnet ___. 2. To ___ is to cause to draw near.
Celsius 1. On this temperature scale water freezes at 0°. 2. On this temperature scale water boils at 100°.	**charge** 1. An electric ___ is the measure of the extra positive or negative particles that an object has. 2. Protons are subatomic particles with a positive ___. Electrons are subatomic particles with a negative___.
circuit(s) 1. Electricity travels in closed loops, called ___. If a ___ is open, the electrons cannot flow. 2. When you flip on a light switch, you close the ___ so the electrons can flow. When you flip it off, you open the ___.	**compound** 1. A ___ is a molecule that contains more than one element. 2. Oxygen (O_2) is *not* a ___ because it contains only one element. Water (H_2O) *is* a ___ because it contains two different elements.
condensation 1. When liquid drops form from water vapor, ___ has occurred. 2. The change from a gaseous state to a liquid state is called ___.	**conduction** 1. The direct transfer of heat from one object to another is called ___. 2. A cold spoon placed in a cup of hot soup will become warmer because of ___.
conductors 1. Most metals are good ___ of electricity; most nonmetals are not. Most metals are also good heat ___. 2. Materials through which an electrical current can pass easily are electrical ___.	**current** 1. The flow of electrons is called an electric ___. 2. An alternating electric ___ is one that reverses its direction at regularly recurring intervals.

Matter and Energy Bingo

© Barbara M. Peller

dissolve(d) 1. ___ matter in a solution is called the solute. 2. A solution is a mixture of one or more solutes that have been ___ in a solvent.	**electricity** 1. Lightning is a naturally occurring form of ___. 2. Static ___ is produced by friction. Current ___ flows through a circuit.
electromagnet 1. An ___ is created by surrounding an iron or steel core with a coil of wire and passing an electric current through the wire. 2. An ___ is magnetized only when an electric current passes through the wire that surrounds the core.	**electrons** 1. The nucleus of an atom is composed of protons and neutrons. ___ orbit the nucleus. 2. ___ are subatomic particles with a negative charge.
element(s) 1. A substance that is made up of only one kind of atom is called an ___. 2. Oxygen, hydrogen, gold and silver are all ___.	**energy** 1. ___ is the ability to cause changes in matter. It cannot be created or destroyed. 2. Some forms of ___ are heat, light, sound, nuclear and mechanical.
evaporation 1. When a liquid changes into a gas, ___ has occurred. 2. The reverse of condensation is ___.	**Fahrenheit** 1. On this temperature scale water freezes at 32°. 2. On this temperature scale water boils at 212°.
frequency 1. ___ refers to the number of times a wave, such as a sound wave, repeats itself in a given time period. 2. The number of vibrations a wave has in a given period of time is called the ___ of the wave.	**gas** 1. This state of matter has no definite shape or volume. 2. This state of matter assumes both the shape and volume of its container.

gravity 1. This is the force that pulls objects toward one another. 2. This force pulls everything toward the center of Earth.	**heat** 1. ___ is the transfer of thermal energy from one substance to another. 2. ___, or thermal, energy is transferred from one substance to another by a difference in temperature.
insulator 1. A material that is a poor conductor, as of heat or electricity, is called an ___. 2. An ___ is used to protect us from the dangerous effects of electricity.	**kinetic** 1. Energy in motion is called ___ energy. 2. As a ball falls, its energy is converted from potential energy to ___ energy.
liquid 1. This state of matter has definite volume but no definite shape. 2. This state of matter takes the shape of its container; its volume stays the same.	**magnet(s)** 1. A ___ is an object that attracts iron or steel and produces a magnetic field. 2. Like poles of ___ repel and unlike poles attract.
mass 1. The amount of matter in an object is its ___. 2. ___ is the amount of matter contained in an object. Unlike weight, it remains the same at any place without regard to gravity.	**matter** 1. ___ is anything that has mass and takes up space. All ___ is made of atoms and molecules. 2. There are three states of ___: solid, liquid and gas.
mixture 1. A ___ is made up of two or more things that have been combined without being changed. 2. A ___ contains two or more kinds of matter, none of which have been changed.	**molecule** 1. Two or more atoms held together by chemical bonds form a ___. 2. A ___ of ozone is made up of three oxygen atoms. A ___ of water is made up of two hydrogen atoms and one oxygen atom.

Matter and Energy Bingo

neutron(s)	**nucleus**
1. A ____ is a subatomic particle with no charge.	1. The ____ of an atom is made up of protons and neutrons.
2. The nucleus of an atom is made up of protons and ____.	2. Most of an atom's mass comes from the protons and neutrons in its ____.
opaque	**pitch**
1. If something does not let light flow through it, we say it is ____.	1. A high ____ sound corresponds to a high-frequency sound wave.
2. If an object is not transparent or translucent, it is ___.	2. A low ____ sound corresponds to a low-frequency sound wave.
potential	**power**
1. ____ energy is stored energy. It exists because of an object's condition or position.	1. ____ is the rate at which work is done or energy emitted or transferred.
2. As a ball falls, its ____ energy is converted to kinetic energy.	2. A watt is a unit of electrical ____.
prism	**protons**
1. A ____ is a solid object that bends light.	1. ____ are subatomic particles with a positive charge.
2. This transparent optical device can be used to break up light into the colors of the spectrum.	2. The nucleus of an atom is composed of ____ and neutrons.
reflection	**refraction**
1. The bouncing of light off an object is called ____.	1. The bending of light that moves from one type of medium to another is called ____.
2. We see our image in a mirror because of ____.	2. When a light wave passes from one medium to another, its speed changes and ____ occurs.

Matter and Energy Bingo

repel 1. Like poles of a magnet ___. 2. To ___ is to cause to push apart.	**solar** 1. The word ___ means "of, derived from, relating to, or caused by the sun." 2. Complete this analogy: thermal : heat :: ___ : sun
solid 1. This state of matter has both a definite shape and a definite volume. 2. The three states of matter are ___, liquid and gas. Ice is the ___ state of water.	**solution** 1. Particles of each substance in a ___ are mixed evenly and do not settle out. 2. A mixture of one or more solutes dissolved in a solvent is called a ___.
sound 1. Characteristics of ___ include amplitude, or loudness; velocity; wavelength; and frequency. 2. ___ is a series of vibrations that can be heard.	**temperature** 1. ___ is the degree of hotness or coldness. It can be measured with a thermometer. 2 ___ is a measure of the average kinetic energy of the molecules in an object.
translucent 1. ___ describes a substance that allows light to pass through it, but scatters the light. 2. ___ objects allow some light to pass through. An image can be seen through it, but the image is blurry.	**transparent** 1. ___ objects are clear. They allow most of the light to pass through. 2. Clear glass is an example of a ___ object. A clear image can be seen through it.
wavelength 1. The distance from one crest to the next crest or one trough to the next trough in a series of waves is called a ___. 2. The distance between repeating units of a wave of a given frequency is called a ___.	**weight** 1. ___ is a measure of the pull of gravity on an object. 2. ___ is sometimes confused with mass, but mass remains the same without regard to gravity and ___ does not.

Matter and Energy Bingo

Matter and Energy Bingo

Fahrenheit	Celsius	heat	translucent	solar
conductors	frequency	sound	matter	magnet(s)
reflection	opaque		gravity	molecule
temperature	attract	evaporation	weight	insulator
kinetic	wavelength	dissolve(d)	atom(s)	gas

Matter and Energy Bingo

translucent	repel	liquid	nucleus	kinetic
insulator	element(s)	conduction	attract	protons
potential	wavelength		electricity	evaporation
matter	power	opaque	transparent	magnet(s)
gas	sound	dissolve(d)	conductors	atom(s)

Matter and Energy Bingo: Card No. 2

Matter and Energy Bingo

translucent	evaporation	matter	weight	reflection
wavelength	Celsius	circuit(s)	frequency	mixture
attract	sound		prism	charge
opaque	potential	kinetic	element(s)	liquid
atom(s)	dissolve(d)	conductors	transparent	heat

Matter and Energy Bingo

opaque	prism	heat	dissolve(d)	kinetic
mass	element(s)	frequency	nucleus	reflection
gravity	conduction		solar	weight
evaporation	energy	sound	conductors	circuit(s)
atom(s)	gas	neutron(s)	electrons	molecule

Matter and Energy Bingo: Card No. 4

Matter and Energy Bingo

gas	solar	attract	conduction	dissolve(d)
mass	evaporation	circuit(s)	opaque	electromagnet
repel	molecule		Celsius	heat
magnet(s)	prism	Fahrenheit	transparent	electrons
matter	conductors	pitch	electricity	gravity

Matter and Energy Bingo: Card No. 5

© Barbara M. Peller

Matter and Energy Bingo

charge	prism	liquid	repel	molecule
weight	attract	electrons	frequency	reflection
nucleus	circuit(s)		conduction	electricity
conductors	kinetic	transparent	neutron(s)	gravity
insulator	evaporation	Fahrenheit	pitch	heat

Matter and Energy Bingo

Fahrenheit	prism	refraction	electromagnet	matter
insulator	heat	wavelength	Celsius	reflection
liquid	weight		electricity	compound
opaque	element(s)	mass	translucent	potential
dissolve(d)	conductors	transparent	neutron(s)	charge

© Barbara M. Peller

Matter and Energy Bingo

gravity	prism	condensation	weight	compound
mass	repel	nucleus	heat	solar
reflection	protons		molecule	conduction
atom(s)	opaque	translucent	electrons	element(s)
sound	conductors	neutron(s)	attract	insulator

Matter and Energy Bingo

electricity	matter	wavelength	reflection	molecule
electrons	repel	gravity	attract	heat
mixture	Fahrenheit		Celsius	condensation
compound	gas	kinetic	electromagnet	refraction
element(s)	transparent	circuit(s)	translucent	solar

Matter and Energy Bingo

temperature	translucent	conduction	nucleus	pitch
molecule	compound	frequency	Celsius	heat
prism	protons		weight	potential
kinetic	magnet(s)	electrons	transparent	mixture
current	gas	liquid	insulator	gravity

Matter and Energy Bingo

charge	protons	attract	electrons	insulator
condensation	mixture	electromagnet	electricity	frequency
mass	repel		liquid	wavelength
current	reflection	transparent	conductors	translucent
circuit(s)	dissolve(d)	Fahrenheit	neutron(s)	matter

Matter and Energy Bingo: Card No. 11

Matter and Energy Bingo

matter	element(s)	mixture	weight	electricity
wavelength	sound	repel	neutron(s)	mass
Fahrenheit	refraction		molecule	nucleus
dissolve(d)	solar	heat	translucent	Celsius
protons	condensation	prism	circuit(s)	compound

Matter and Energy Bingo: Card No. 12

Matter and Energy Bingo

current	solar	charge	mixture	molecule
repel	condensation	prism	electricity	potential
weight	conduction		wavelength	refraction
gravity	transparent	compound	protons	translucent
conductors	magnet(s)	neutron(s)	Fahrenheit	electromagnet

Matter and Energy Bingo

dissolve(d)	repel	attract	electricity	current
compound	Fahrenheit	mixture	Celsius	potential
electrons	weight		liquid	conduction
magnet(s)	transparent	prism	circuit(s)	charge
conductors	nucleus	protons	insulator	gravity

Matter and Energy Bingo

electromagnet	electricity	attract	matter	heat
charge	pitch	frequency	repel	electrons
molecule	Fahrenheit		reflection	weight
conductors	mixture	condensation	transparent	current
insulator	element(s)	neutron(s)	liquid	wavelength

Matter and Energy Bingo

conduction	solution	condensation	pitch	power
nucleus	protons	refraction	mass	temperature
current	solar		molecule	wavelength
opaque	element(s)	conductors	electromagnet	translucent
electrons	mixture	neutron(s)	compound	potential

Matter and Energy Bingo

current	solid	energy	mixture	conductors
electromagnet	electrons	transparent	weight	refraction
electricity	temperature		solution	condensation
gas	insulator	gravity	attract	potential
kinetic	circuit(s)	matter	translucent	solar

Matter and Energy Bingo

heat	prism	compound	electrons	nucleus
gas	current	attract	molecule	circuit(s)
electricity	potential		energy	pitch
protons	frequency	transparent	temperature	liquid
solution	mixture	kinetic	solid	charge

Matter and Energy Bingo

molecule	charge	mixture	condensation	protons
electromagnet	dissolve(d)	pitch	matter	temperature
solid	weight		Celsius	attract
liquid	solution	kinetic	element(s)	energy
reflection	power	insulator	gravity	neutron(s)

Matter and Energy Bingo

protons	solid	temperature	mixture	Celsius
conduction	wavelength	mass	kinetic	nucleus
solar	refraction		opaque	energy
gas	gravity	atom(s)	element(s)	solution
evaporation	sound	power	translucent	frequency

Matter and Energy Bingo: Card No. 20

Matter and Energy Bingo

charge	gas	mass	mixture	magnet(s)
solar	energy	compound	condensation	Fahrenheit
potential	insulator		solid	attract
kinetic	matter	solution	electromagnet	gravity
opaque	power	neutron(s)	current	element(s)

Matter and Energy Bingo: Card No. 21

Matter and Energy Bingo

reflection	liquid	energy	repel	current
nucleus	temperature	heat	condensation	Celsius
compound	weight		Fahrenheit	refraction
solution	gas	element(s)	frequency	dissolve(d)
power	circuit(s)	solid	potential	mass

© Barbara M. Peller

Matter and Energy Bingo

conduction	solid	matter	repel	neutron(s)
charge	protons	insulator	electromagnet	frequency
liquid	current		atom(s)	Fahrenheit
potential	sound	solution	circuit(s)	element(s)
magnet(s)	gravity	power	kinetic	energy

© Barbara M. Peller

Matter and Energy Bingo

conduction	protons	dissolve(d)	solid	condensation
molecule	neutron(s)	mass	nucleus	Fahrenheit
refraction	pitch		current	potential
magnet(s)	atom(s)	solution	circuit(s)	solar
evaporation	opaque	power	temperature	sound

Matter and Energy Bingo

opaque	mass	solid	attract	energy
frequency	magnet(s)	electromagnet	conduction	Celsius
solar	condensation		atom(s)	solution
pitch	gas	sound	power	temperature
neutron(s)	dissolve(d)	compound	electrons	evaporation

Matter and Energy Bingo: Card No. 25

Matter and Energy Bingo

energy	solid	atom(s)	nucleus	pitch
liquid	weight	condensation	protons	conduction
magnet(s)	kinetic		temperature	opaque
current	repel	gas	power	solution
refraction	electrons	attract	sound	evaporation

© Barbara M. Peller

Matter and Energy Bingo

atom(s)	compound	solid	protons	wavelength
magnet(s)	liquid	electromagnet	solution	Celsius
transparent	sound		power	opaque
pitch	charge	evaporation	mass	frequency
current	temperature	energy	reflection	refraction

Matter and Energy Bingo

molecule	prism	translucent	solid	compound
wavelength	energy	atom(s)	kinetic	temperature
sound	potential		pitch	nucleus
refraction	reflection	insulator	power	solution
repel	electricity	current	evaporation	magnet(s)

Matter and Energy Bingo

energy	prism	pitch	electromagnet	electricity
magnet(s)	kinetic	mass	refraction	reflection
solar	solid		Celsius	atom(s)
wavelength	gas	heat	power	solution
conduction	condensation	evaporation	charge	sound

Matter and Energy Bingo

dissolve(d)	solid	nucleus	electricity	solution
frequency	pitch	liquid	temperature	Celsius
evaporation	sound		refraction	mass
magnet(s)	charge	energy	power	atom(s)
gas	matter	circuit(s)	prism	heat

www.ingramcontent.com/pod-product-compliance
Lightning Source LLC
Chambersburg PA
CBHW051419200326
41520CB00023B/7297